GOLD PANNING FOR PROFIT

by

WALTER J. ROBERTSON
MINING ENGINEER

— 14th Edition —

A nugget from Talkeetna, Alaska, worth from 200 to 300 times the silver dollar behind it.

Contents

CHAPTER I

SHALL WE GO GOLD PANNING?

A recent publication on the subject of placer gold mining included the following warning in its preface:

> "A person not in robust health or one who has not sufficient funds to finance his entire trip runs a splendid chance of starving to death if he tackles placer mining. . ."

Although that particular book referred primarily to one of our Southwestern states, other books in recent years have carried similar warnings. It would seem that on one hand the authors encourage prospectors by writing of the fascination of gold placering, and with the other hand attempt to restrain the adventurous beginner by emphasizing more or less real dangers.

Frankly, I believe that too much emphasis has been placed upon risks placer miners have to take. After all, practically every enterprise has the possibility of failure, and placer mining is no exception. As a matter of fact, many of the adventurers who flocked to California in 1848 and 1849, and during the years immediately following, were unable to make their living even in the rich, virgin California gold fields. Briefly, you may be successful — and maybe you won't — but win, lose or draw, you will have a wonderful time doing it!

But let us be realists. If you have a good steady job, and if you're more or less satisfied with your pay check, I would not recommend that you immediately take up gold placering as a vocation. If you have others who are financially dependent upon you, don't chuck up your job and set out for the hills. And if you cannot stand the isolation of the hills and deserts, it may be you're just not cut out to be a good gold placer man. No, I'm not telling you to forget all about it! Far from that! Instead, I suggest you first try gold panning as an avocation or hobby. After all, you wouldn't attempt a cross-country motor trip without first learning how to drive the car. First you would take very short drives and gradually go further afield as you develop confidence. Similarly, I recommend that you first take short prospecting trips on weekends, during your vacations, or in between jobs. You'll learn a lot about placer mining in this way and you will also learn whether or not you are temperamentally and physically suited to the work.

This booklet does not profess to attempt to teach you all there is to learn about gold panning, or gold placering, to use the more general term. Neither will reading this booklet make a finished prospector of you. After you have read this book once — and preferably, several times — you should have a pretty fair idea as to what it is all about. If you are cut out to be a live-wire gold panner, you will undoubtedly want to do much more reading on the subject. To aid you in that respect, I have prepared a bibliography which you will find at the end of this book. You can learn something from each of these references, and the more of them you read, the better you will be prepared to take up the adventurous role of gold placering.

Because of lack of time, money, physical infirmities, or any one of a dozen reasons, some of my readers will not be able to go out in the field to engage in the absorbing task of gold panning. These folks will become expert armchair placer miners! But to become genuinely proficient, you will have to obtain a lot of experience by actually going out into the deserts or up into the hills. No amount of reading can take the place of real experience, but the combination of studying good books and the diligent application of such study while out in the placer country is an unbeatable one. Such a combination, plus a reasonable amount of luck, can pay you rich dividends.

At this point, it might not be amiss to point out that luck has often seemed to play an important part in small scale gold placering. Just the other day, an article appeared in the papers to the effect that a truck driver for one of the local sand and gravel plants found a gold nugget worth $400. While waiting to have his truck loaded, he walked just a few feet away to a dry creek bed. He noticed the nugget lying at his feet and merely had to bend over and pick it up! He didn't know the first thing about gold placering, but still he found a beautiful and valuable nugget.

Similarly, several years ago a local young couple were on their way to spend their vacation up in the hills near here. They stopped their car at a spot not over forty miles from Stockton (Calif.) and near a small stream. While they stood at the edge of the stream, the lady noticed something shining in the water at her feet. Upon picking it out of the water, she found it was a nugget of placer gold. Subsequent weighing showed it to be worth about $125. They spent the balance of their vacation searching in the same general area for more nuggets — and didn't find a thing! What does it prove? That there was no more gold to be found? No, I don't think so. I feel it tends to prove that if you are in a good placer gold country, even the rankest beginner might find some gold through sheer luck. And I think it also goes to prove that without a fundamental knowledge of gold placering one might spend weeks without finding more than a color or two.

CHAPTER II

WHAT OTHER PROSPECTORS ARE DOING

The items contained in this chapter have been taken from recent issues of various mining publications, and are presented here to give an idea as to what some prospectors are accomplishing today. In the California Mining Journal for October we read:

> Lester Seepie, Indian, who has been placer mining in the Tule Canyon district, some 50 miles or so west of Goldfield (Nevada), in Tonopah displayed a bottle of small gold nuggets which he had recovered in eight days. The gold — three ounces and two pennyweights — ranged in size from coarse grains to small nuggets. Seepie at times uses a dry washer, but the gold he showed was recovered in a rocker with water. Some time back he 'rocked' out six wheelbarrows of gravel from which he recovered two ounces and six pennyweights of gold.

In the November issue of the same magazine we learn:

> Mining of profitable gravel is reported proceeding steadily at the Morris Ravine Mine (Butte County, Calif.) under the management of J. H. Sharpe. The property is said to have produced gold continuously for 50 years. During the late war when placer mines were allowed to take out only 100 cu. yds. per month, the mine had a monthly production of $3,500.

The December issue of the California Mining Journal carried an item which shows that rich gold bearing material is still being found. It reported as follows:

> A gold strike near the Alaska-Yukon border wherein the crushed ore is reputedly worth $100 per pan is reported in northern newspapers as we go to press. Alphonso Champagne, veteran northern prospector, started Peace River towns buzzing when he walked in on some business men at Fort St. John and plunked down a jar containing gold-studded quartz.

Further on, the same article told of another strike as follows:

> Another rich gold strike was reported to have been made within 80 miles of Aklavik by two prospectors during the month of August. The ore samples were said to have been so rich that the prospectors kept the samples with them day and night and would allow no one to handle them.

The following item from *The Mining Record* of July 10 reads almost like fiction:

> His hair was long and unkempt, his beard wild and bushy, this stranger from the creeks who walked into the Pioneer Hotel (Fairbanks, Alaska) recently packing a well filled flour sack on his shoulder. He looked like Dan McGrew out of a Service poem. Said Manager Geo. Gilbertson as the stranger laid the sack down on the counter, "What you got there?" Replied the man from the creeks, "It's our first cleanup of the season — 1,000 ounces of gold." He registered. It was Jim Pierce, who with Jim McDonald and Charley Cravy, is mining in the Manley Hot Springs district. He later returned to the camp via Lon Brennan plane.

Some interesting phases of placer mining are related in the following item reported in the October issue of the *California Mining Journal*:

> H. C. Windell and Lawrence Nessler are developing the Monte Cristo drift placer mine, located 13 miles north of Downieville (Calif.), and 13 miles off a road. The operators are required to walk in that 13 miles, not even having a trail when they began work. They have driven their tunnel several hundred feet into a secondary channel and are approaching what they believe to be the primary deposit. In the old days miners enjoyed working rich diggin's in what is said to be a spill-over from the primary source of the riches. Windell estimates they can make an ounce per day with hand methods and have made several tests which average $28 per yard. Former Monte Cristo miners are said to have taken out wheelbarrows of gravel that ran up to $125 and produced nuggets that ran from a half ounce up to $125.

There has been a tendency in recent years for some writers to say that all worthwhile gold placer ground has already been discovered and located, and that the beginner placer miner has very little chance. Such defeatists should read the following authentic account printed in the October 2 issue of *The Mining Record*:

> Gold deposits assaying as high as $775 a ton have been discovered in the Rainbow Valley district in Natrona County (Wyoming), H. P. Waite, president of the Nevada Wyoming Company, of Casper, reported last Friday.
>
> After a conference here with Dr. H. G. Fisk, a director of the Natural Resources Institute at the University of Wyoming, Waite said the gravel deposit, estimated in excess of 36 million tons, also contained silver, mercury, and platinum. He said mining operations by the company would begin in two weeks.

From Fairbanks, Alaska, we learn of several recent and very interesting cleanups:

> The Burwash Mining Co., whose ground is about 187 miles from Whitehorse, had a recent cleanup of 343 ounces of placer gold, valued at approximately $12,000, the result of a 20-day shift working 10 hours a day.
>
> Two previous cleanups taken out this season, according to the *Whitehorse Star,* represented 96 and 122 ounces and it is anticipated that a further two pannings will be secured before the season ends. Last year, the first of its operation, during a short season the company took out about 400 ounces.

Those of my readers who are World War II veterans will be interested in the following reported from Alturas (Calif.) in the July 10 issue of *The Mining Record*:

> Snow is melting rapidly in the high hills of the High Grade district at the northeast corner of Modoc County (Calif.) and a gold rush into the area is anticipated as soon as roads have been opened. Wellman Smith, a World War II veteran, discovered rich gold ore on his Moonlight claims last fall but deep snow halted mining soon afterward.
>
> Numerous claims were located in the area by prospectors. Smith says he received a number of offers for his claims. Options are said to have been taken on several properties by representatives of major mining companies. Ore discovered by Smith is said to average close to $500 a ton.

The Lovelock Miner-Review (Nevada) recently reported the following interesting item:

> S. L Gordon, has two miles of Willow Creek Canyon which during the past 15 years has produced some of Nevada's largest gold nuggets. After sampling the ground Gordon has completed a reservoir at the head of his ground and fitted his dam with an automatic ground sluicer, or 6' x 6' "shooter." When the water supply arrives this winter and rises against the automatic gate, the gate will be tripped and dump a large quantity of water on the gravel below. Force will be given by the water tearing down a 30% grade, the values being recovered in sluice boxes below. A new rain storm will bring another water supply and the operation repeated. In between times Gordon can make his cleanups.

In a lighter vein comes the following little item from Dallas, Ore.:

> Phil Mueller would like to know where those mallards he shot on the Little Luckiamute River near the end of the season come from. In their gizzards were several tiny gold nuggets and gold flakes!

The California Mining Journal reports a somewhat similar item in its February issue:

> Back in 1931 as the last Gold Rush was starting, the late Horace Onyett of Oroville, Calif., built California's first doodlebug dredge at Palermo, Butte County. He was too busy to test his ground but did examine the dirt from a squirrel hole. He found gold in the squirrel tailings!

Finally, I guess we all wish we could do our gold placer mining as conveniently as those folks who live in Oroville, California. *The California Mining Journal* recently related:

> Oroville residents were recovering from a one-day "gold rush" when several "miners" recovered nearly an ounce of gold nuggets in a downtown gutter.
>
> Best haul of the day was made by J. J. Beck who first spotted the nuggets through the rain water that had collected in the gutter. He scooped up a handful of sand and gravel which, when washed, produced nearly a half ounce of coarse gold.
>
> Forest Olson, 25, of Minnesota, found a crevice in the curbing nearby that produced several nuggets and considerable fine gold. Leo Erwin, 31, a local resident, discovered gold at another point 25 yards away.

CHAPTER III

SOME FACTS ABOUT GOLD

Now that you've decided to go gold panning, it's high time you knew what you're looking for. When you find those shining particles of gold in your pan or sluice box, you'll know they're GOLD. But it won't hurt to fill in here with some interesting general facts.

While pure gold has a characteristic brass-yellow color, most placer gold contains sufficient alloying material, such as silver or copper, to modify its color. For instance, most placer gold contains some silver which tends to lighten the color. If the gold contains a very high percentage of silver, the gold tends to assume a silver-white color with a yellowish tint.

Based upon careful assays of a large quantity of placer gold derived from a majority of the placer gold producing areas in this country, it has been found that the gold's fineness will roughly average 895. In other words, in 1,000 parts of the material as recovered, about 895 parts will be pure gold, and approximately 105 parts will be silver, copper, and minor amounts of base material.

Gold is a very heavy metal. Expressed in more technical terms, pure gold has a specific gravity of about 19.3 and weighs approximately 1,187 pounds per cubic foot. Without going into needless detail, it is the high specific gravity of gold which causes it to settle in the bottom of your pan, or to be caught in the riffles of your sluice, instead of being carried away with the lighter rocks and gravel by the water. The hardness of gold ranges between 2.5 and 3.

If you're still not certain those golden particles in your pan are really gold, try pounding a piece on an anvil with a hammer. If it flattens without breaking or shattering, you can be pretty sure it's the real thing. Iron pyrites, sometimes called "fools' gold," will crumble and powder when so pounded. Similarly, if your knife blade cuts or indents the metallic particle (as the blade would cut or mark metallic lead), you're entitled to yell, "Eureka!" (which I understand is the Greek equivalent of "I have found it!"). It is interesting to note that gold is the most malleable of all metals. In fact, gold can be hammered in sheets so thin that it would take 280,000 of them to make a pile only one inch high! As a matter of fact, gold can be beaten even thinner and a thickness of 1/367,500-inch has been attained. One grain of gold, worth about 7c, and of this thickness, would cover 52 square inches.

Because of its high ductility, gold can be made into wire so fine that 50 miles of it would weigh only one ounce!

In the course of your placering, you'll come across gold particles in a wide range of sizes. For instance, gold so fine that it takes nearly 2,000 particles thereof to be worth about one cent can be recognized in your pan. Recovery of such fine gold (often called flour gold) is pretty difficult and I advise against wasting time with it. On the other hand, some of the nuggets may run pretty large. One of the largest nuggets ever found was the famous "Welcome Nugget," which was found in Australia in 1858, and weighed over ONE HUNDRED AND FIFTY POUNDS! Another large nugget that weighed 2,440 troy ounces was found near Carson Hill (California) in 1854. Up in the Klondike, a nugget weighing 85 ounces is the largest found so far, to the best of my knowledge. More recently, a nugget weighing 21 pounds and another of 31½ pounds were found in the Ural Mountains, in Russia.

At the Barbarossa Placer, a dry diggin's at an elevation of 4,000 feet in Arizona, a nugget was produced that weighed 22 ounces. Shortly after the discovery of the Klondike District (in Esmeralda County, Nevada), a nugget valued at $1,200 was found. And one of the largest nuggets found in Nevada weighed 24 pounds and was found in the Osceola District!

Just a few weeks ago, a very nice nugget was found about eight miles west of Mountain City, Nevada. The prospector was operating a small sluice although the nugget was actually picked up in the "puddle box" which was being used to break up the material before running it through the sluice. As reported, the nugget was as big around as a dollar and almost a half-inch thick. It weighed four and one-half ounces, and, with gold at $35 per ounce, it was worth about $157.00! Well worth bending over to pick up, I'd say!

I have cited only a few cases where really nice big nuggets were found. There are hundreds more that might be mentioned, if space permitted. But any way one looks at it, all the nuggets certainly have not been found. Other prospectors will find some big ones — you can, too!

CHAPTER IV

WHERE TO PROSPECT

In the words of a formerly popular radio program, "That's the $64 question!" Another slant on the matter of where to look for placer gold may be found in the words of the old timer who, when asked that question, said, "Where it be's; there it be's," and then sadly added, "and there ben't I."

However, the best general advice is to prospect for placer gold in localities where such gold has already been found. In the back of this book you will find maps advertised that will show you exactly where placer gold has already been mined. I might add that these maps would have been reproduced herein were it not impossible to show sufficient detail on this page size. Although some of the areas shown on the maps have been worked extensively, it is improbable that even the beginner will not be able to show at least some very interesting colors at practically any of the areas shown on the maps. A few years ago a nice $152 nugget was found lying in a waste dump in the Weaver District in Arizona! No, the old timers didn't find it all!

Naturally, it would be impossible to give details of all placer mining areas in a single book. Even the maps referred to above, although excellent and the very best available, do not show you the precise spot where to take your first panful of gold-bearing gravel. In other words, you still will have to do a certain amount of prospecting. After you have selected a specific area in which to prospect, you may write to the Mining Bureau or the Department of Natural Resources of that particular state for more detailed information. The various state agencies are always willing to help the prospector in every way they can. At least, that has been my experience.

If you care to, you may write to me for more specific information regarding likely spots to prospect. It has always been my policy to help the new prospector in every way I possibly can. I cannot guarantee that I can help you in each and every instance, but I promise to do my best. However, if you write, please be sure to enclose a stamped self-addressed envelope with your inquiry. I'll try to answer your questions as soon as possible. If I happen to be away on a prospecting trip, and I am gone frequently, there may be a little delay in my reply. But you can rest assured that your letter will be answered.

I have noticed a tendency for beginner prospectors to confine their efforts to not more than a few hundred yards from a paved highway. I won't say that some gold might not be found that way, but I do say your chances of making a rich strike would be mighty slim. Bear in mind, the easier it is to reach a given spot, the greater the likelihood that spot has been worked over many times before. So my advice is to travel up the less accessible streams and creeks which empty into the principal rivers. When you get far enough away from the roads, you will have greatly increased your chances of locating some really rich placer ground.

CHAPTER V

GOLD PLACERS IN GENERAL

Gold placers are deposits of gravel and sand which contain appreciable concentrations of gold. In general, the formation of such gold placers has resulted from the natural erosion of pre-existing rocks of more or less gold content. It is interesting to note that many rich placer deposits have been formed by the erosion of gold-bearing veins or lodes which, in many cases, were of rather low grade.

A technical discussion of all the geological factors involved in the formation of gold placers is beyond the scope of this handbook. The bibliography contains several excellent references to which the reader is referred should a more detailed discussion be desired.

The professional geologist classifies placers into several types, such as stream placers, eluvial placers, bajada or desert placers, glacial-stream placers, eolian placers, and beach placers. However, the average small operator will find his best bet in surface stream placers of various sorts.

A stream placer is one in which the gold has been transported by water to its present location. Since the deposition of gold in many instances has taken place tens of thousands of years ago, the location of the old stream may have no apparent relation to existing water courses. Hence, placer deposits may be found high up on the hillsides or as much as several hundred feet below the present surface. Frequently, rich bench or terrace placers are overlooked just because they are fifty, a hundred, or more feet higher than the present nearby stream. However, such deposits are not always overlooked. A Mexican, looking for strayed animals, discovered gold nuggets on the top of Rich Hill in southern Yavapai County, Arizona. This discovery soon led to the finding of the rich Weaver and Antelope Creek placers.

Placer gold is ordinarily found pretty close to bed rock. This is due to the fact that gold is six (or more) times heavier than the rock with which it is associated and thus tends to sink and concentrate along the bottom or bed rock. Naturally, the coarser pieces of gold settle first, whereas the finer flakes of gold are carried along much farther. Where the bed rock is fractured or broken, the gold particles will concentrate in such crevices, sometimes to a depth of five or more feet. Such cracks might be considered Nature's own riffles and sometimes they are marvelously rich.

In prospecting for placer gold, surface pannings might be very poor whereas if the surface material (called over-burden) is cleared away first, the pannings nearer bed rock mght go very much higher. Values of fifty or more dollars per cubic yard for the material immediately above bed rock are not too unusual. Hard and fast rules are difficult to state, but you won't go far wrong to assume that if you get pretty good colors (small pieces of gold) over a given surface area, you'll do a lot better by going down to or near bed rock. Another thing, no matter where you prospect, take the time to get acquainted with some of the old-timers in the locality. Casual conversations will not only help to familiarize you with the district, but may provide a tip or two that may save you a lot of time — and possibly money. Gold panning always is an interesting subject, and most folks in a placer country are willing to talk about it to your heart's content!

Remember that I said the coarser gold sinks first whereas the finer material is carried considerably farther. If your gold is very fine and flaky, you might do well to prospect further up the stream and search for the coarser and heavier gold, and maybe some "hefty" nuggets!

Another thing, in the case of existing creeks and streams, watch most carefully for more or less abrupt changes in their direction. Whenever there is a slowing down of the rate of flow of a stream, such slowing down may have been sufficient to cause the stream to drop or deposit the gold it was carrying. Naturally, such spots are well worth prospecting. Suppose you are walking along a stream and it makes a fairly sharp turn. That part of the stream on the inside of the curve will flow more slowly than the part of the stream on the outside of the curve. Hence, carefully prospect that part of the stream bed or bank on the inside of the curve. You stand your best chance of finding gold there.

The appearance of large amounts of black sands in your pan or sluice is a very favorable indication. It does not necessarily prove that gold will also be found since black sand, to a greater or lesser degree, can be found almost anywhere. However, since black sands are composed of minerals of high specific gravities, conditions that were favorable for their deposition would also have been favorable for the deposition of gold and for the formation of placer gold deposits.

CHAPTER VI

INTRODUCTION TO PANNING

We are all familiar with pictures of a bewhiskered prospector kneeling at the edge of a small creek and holding in his hands the traditional miners' gold washing pan. The pan is just as important to the modern prospector as it was to the Argonauts of 1848 and 1849.

The pan is usually made of heavy-gauge steel (sometimes of copper, aluminum, etc.) and varies between twelve and eighteen inches in diameter. The pan is between two and two and one-half inches in depth, with sides having a slope of around thirty to forty degrees.

The use of the pan is much easier to demonstrate than to describe. When you go into placer mining country, you will have lots of chances to watch how panning is performed. Watch the old-timers — they know their business. The technique of panning is well described in a recent publication of the State of California as follows:

> "The object of panning is to concentrate the heavier materials by washing away the lighter. To do this efficiently, all material should be as even a size as possible. The pan is filled about three-quarters full of gravel to be washed, then it is submerged in water. First, the large gravel is picked out by hand, then the clay is broken up, after which the operator raises the pan to the edge of the water, inclining it slightly away from him, moving it with a circular motion combined with a slight jerk, thus stirring up the mud and light sand and allowing it to float off.
>
> "This is continued until only the heavier materials remain, such as the gold, black sand, and other materials having a high specific gravity. These concentrates are saved until a large quantity accumulates, after which the gold is separated from them. It may be picked out by hand, amalgamated with quicksilver (sometimes in a copper-bottomed pan,) or in some cases where the separation is extremely difficult, and the quality and quantity justifies, they are shipped to a smelter. Panning may be best learned by watching an old-timer or experienced operator at work, learning certain tricks of the trade from him."

But don't worry about learning to use the pan. It is just a matter of practice. I recommend that you first practice panning in your back yard. Fill your pan about three-fourths full of dry sand and gravel. Now mix in about a tablespoonful of lead shot. (Even better would be about a half teaspoonful of the black sand concentrates and gold you will find advertised in the back of this booklet.) Fill a large tub with water, kneel down with your pan, and go to it. You'll find that with practice you will be able to pan away practically all the rock and sand, and to save all of the lead shot — or the black sand concentrates. When you can recover all (or nearly all) of the lead shot or the black sand, you will be able to do a creditable job of actual gold panning.

You will find that you can pan between one-half and one cubic yard in about ten hours. This represents between 100 and 200 regulation size pansful. It follows, therefore, that the placer gravel will have to be fairly rich in order to give even moderate financial returns when the gold is to be recovered by hand panning alone.

CHAPTER VII

HOW ABOUT SLUICING?

Before you finish your first full day of panning, your aching muscles may have you wondering if there isn't an easier way to recover placer gold. There is, provided you have an adequate supply of water. For instance, there is the rocker or cradle, but you will find it nearly as much work as panning and the amount of material handled in a given time is increased but slightly. The rocker uses quite a bit less water than a sluice, but unless the gravel is rich, and water is scare, my personal advice is to skip the rocker. In all fairness, however, I should mention that many placer miners have used a rocker very profitably. For illustration, in the Limerick Canyon placer (Nevada,) several small operators who used only rockers and hauled water for six miles, netted over $40 a day in one 60-day season. A few patches of this gravel ran to $5.00 per pan! You will find that some sort of sluice will enable you to handle a maximum of placer material with minimum effort and at minimum cost.

Sluicing is a method of separating and recovering gold from the placer gravel by the use of a sluice and running water. The term sluice may be applied to any sloping trough or ditch through which the placer gravel is washed by the action of running water. The gold is caught or trapped by what are known as riffles. Riffles are obstructions along the bottom of the sluice and which retard the movement of the gold and heavy minerals, and which form pockets to trap the gold.

Fig. 2 illustrates the construction of what is known as a DIP BOX. This is a modification of the standard sluice and is commonly used where water is somewhat scarce and there is insufficient grade for a regular sluice. The operation is simplicity itself. Set up the dip box with a grade of from six to twelve inches in six or eight feet. The upper end of the dip box should be about waist high. Dump a small bucketful of gold-bearing sand and gravel on the upper end of the box. Water from a dipper, bucket, or hose is poured over the gravel, thus breaking it up and washing it down the dip box to discharge the barren material over the lower end. The gold will be caught in the screen and burlap, or whatever material you use to trap the gold. Large rocks will have to be thrown out by hand.

The dip box is a very nice and efficient outfit, particularly when one rigs up a small gasoline engine operated pump to pump the neces-

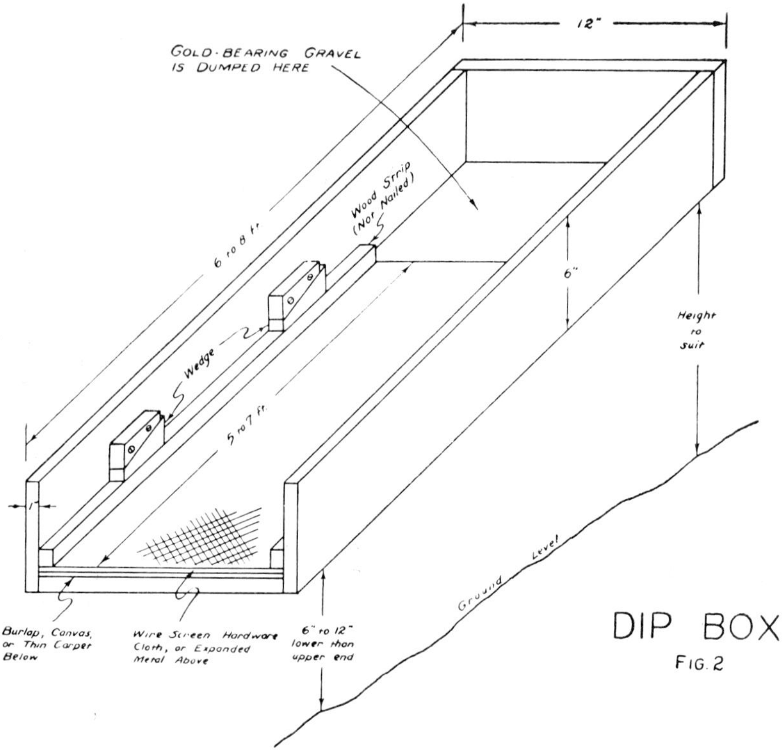

DIP BOX

FIG. 2

sary water from a nearby creek or pond and to discharge the water in a continuous stream at the head of the dip box. A continuous regulated stream of water will prove considerably more efficient than the intermittent flow provided by a dipper or bucket.

Figure 3 illustrates a LONG TOM, another modification of the ordinary sluice. The gold-bearing gravel is shoveled into the head of the long tom and is washed by water entering at the same point. After shoveling in the material, work it around with the shovel or a fork. This facilitates the breaking up of the material, particularly if it is partially cemented. The finer material passes through the screen at the lower end of the second section. Riffles in the third section are designed to catch and retain the gold. The larger rocks which cannot pass through the screen are thrown out by hand. The long tom will keep two to four men busy, and has a capacity of from four to six cubic yards per man per ten hour day.

A regular sluice box layout is shown in Figure 4. The more or less standard length of a section of sluice box is twelve feet. Frequently, additional sections are placed either end to end, or telescoping one in-

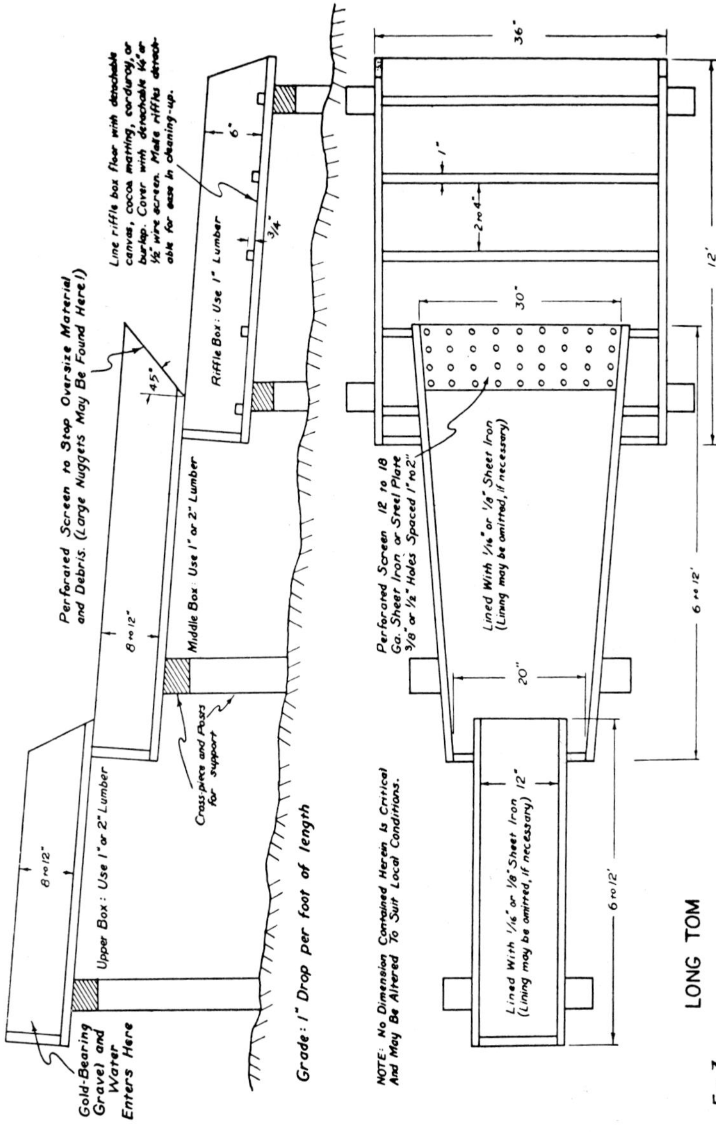

Fig. 3

to the other. Whether one, two, three, or more sections should be used is a matter of simple economics. If the gold clean-up from a given section of sluice is worth more than the cost of its materials, labor, and maintenance, that particular section is justified from a gold recovery standpoint. It should be pointed out, however, that additional sections are frequently required in order to carry the tailings, or barren gravel, to a convenient dumping spot. These additional sections are necessary despite the fact they may save little or no gold.

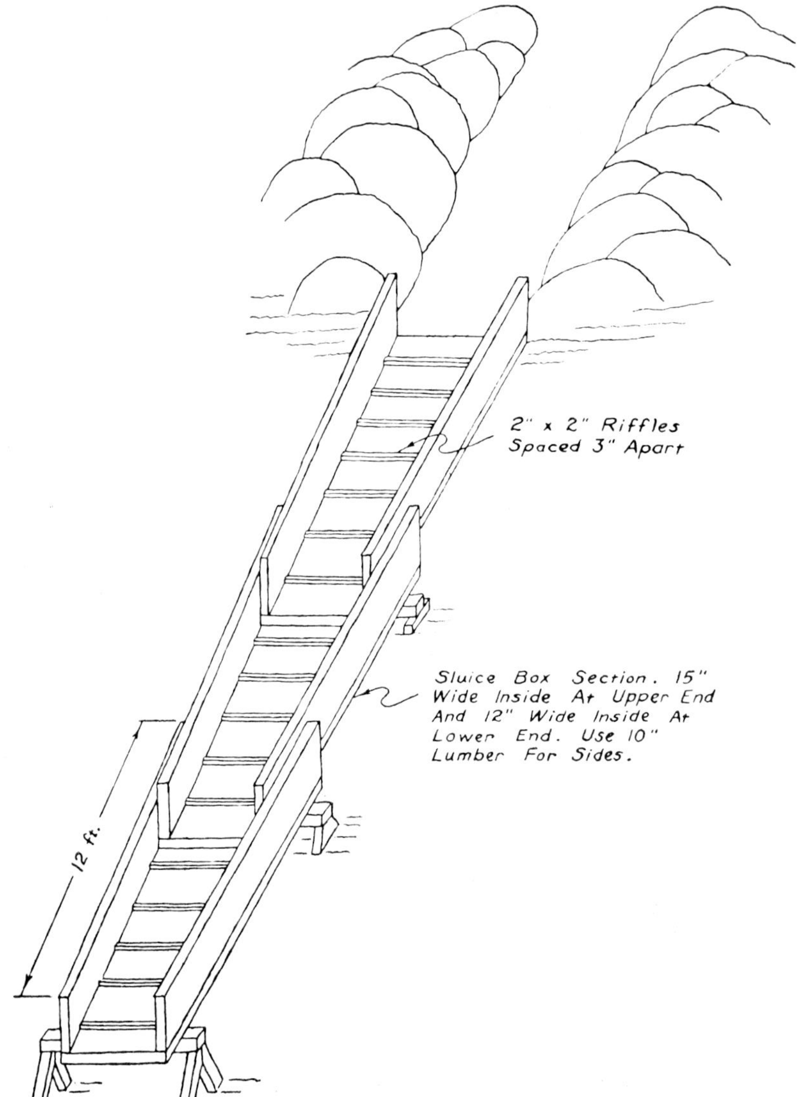

FIG. 4
SIMPLE SLUICE BOX LAYOUT

I believe that the average small placer operator will find that three or four 12-foot sections of sluice box will prove to be a good all around arrangement. This is not a hard and fast rule — it may be found that as much as one hundred or more feet may prove profitable.

Sluice boxes may be made of 1-inch or 2-inch lumber, either 10, 12, 14, or 16-inches wide, and either rough or smooth. It is advantageous to use surfaced lumber since the sluice not only can be made more nearly watertight, but less loss of fine gold at the joints will be experienced. The sluice box should be set at a grade of from 1 in 20 to 1 in 24. A rather flat grade reduces the duty of the water, thus setting a rather low limit on the size of rocks which will be carried by a given flow of water — but may help to save the very fine gold. Conversely, a steep grade (say over 1 in 10,) is rarely used since all but the very coarsest gold would be carried away by the force of the water. In general, determining the proper grade for your sluice is largely a matter of experimentation. However, here's a tip that will help your gold recovery: Every second or third sluice, either increase or decrease the grade of the section slightly. This tends to break up and modify the flow of the placer gravel, with a resultant improvement in gold recovery. After you have operated your sluice for a while, take time out to pan some of the tailings discharged over the end of the last sluice box. If panning shows very little or no gold, the sluice is functioning properly. If considerable gold is found in the tailings, it's your signal to correct the sluice operation. This may be effected by modifying the amount of water or gravel going through the sluice, by changing the grade of the sluice or by another riffle arrangement, or by a combination of these factors.

Don't make your sluice box less than 10 inches wide or its carrying capacity will be too small. On the other hand, be careful not to go to the opposite extreme of making the boxes too wide, or you may experience difficulty with clogging. If the box is too wide for a given amount of water, the depth and velocity of water will be insufficient to properly carry the gold-bearing gravel through the sluice.

CHAPTER VIII

USING QUICKSILVER

When properly used, quicksilver (mercury) can be a lot of help in recovering placer gold. When improperly used, it can be a nuisance! Its fluidity will permit the 'quick' (a colloquial term for quicksilver) to escape and be lost through the slightest crack in your sluice. Its vapors are extremely poisonous if inhaled; in fact, contact with the mercury with bare hands should be avoided as much as possible. Don't let the foregoing scare you — just use the quicksilver as recommended and you need not expect any trouble.

Quicksilver has three characteristics which makes its use desirable in many placer operations:

(1) Its ability to amalgamate with gold (and silver.)

(2) Its very high specific gravity which permits it to lie quietly under a moving stream of water and placer gravel.

(3) Its boiling point of around 675° F. which permits it to be driven off by heat from the gold with which it has amalgamated.

The amalgamation referred to above is simply the alloying of the mercury and the gold. If you take a globule of mercury and a pinch of fine gold dust and shake them together, the gold disappears into the mercury. The mercury will continue to amalgamate with additional gold until the mercury is no longer fluid, but thick and pasty. Roughly, this amalgam will then run about one-third by weight of gold and two-thirds by weight of quicksilver.

In ordinary panning, there is little, if any, merit in the use of quicksilver. It is highly probable that the quicksilver losses would exceed the value of the fine gold saved. However, in the panning of concentrates, the use of quicksilver is definitely advantageous. The term concentrate is used in referring to the residual one or two teaspoonsful of material left after panning a full pan of gravel, and in connection with the concentrated material obtained in cleaning up the dip box, rocker, long tom, sluice, etc.

The ordinary steel pan can be used in cleaning up concentrates by first adding about a quarter to a half teaspoonful of mercury. Then add about five pounds of your panning or sluice box concentrates. The use of more mercury will do no harm. Place the pan under water and agitate until you can no longer see any free or uncombined gold. This operation is best performed in a tub of water which will catch any quicksilver or gold which might otherwise accidentally spill out of the pan and be lost. Now pan in the regular way until only the free mercury and amalgam remain — this can be poured into a wide mouth bottle for further treatment.

Copper plated pans, or steel pans with copper bottoms, are sometimes used to recover the fine gold in concentrates. To amalgamate the copper bottom of a pan, first polish the inside to a shiny brightness. Now pour a little mercury in the pan and rub it in with a clean cloth. The mercury will amalgamate with the copper to give the latter a bright silvery appearance. Before using the pan to clean concentrates, first add a few extra drops of mercury. You will find that the amalgamated surface quickly picks up and holds practically all of the fine free gold in the concentrates. The amalgam thus formed can be conveniently scraped off with a putty knife. From time to time, additional quicksilver is added as required.

I have found it very handy to use the following plan in cleaning up small batches of concentrates, particularly when the concentrates contain a considerable amount of so-called "rusty gold" which amalgamates only with difficulty. Take an ordinary half-gallon fruit jar and add a teaspoonful of mercury. Then fill the jar about half full of concentrates. Add about a pint of water and a teaspoonful of ordinary household lye. Shake the jar vigorously for several minutes to thoroughly mix the contents. Then, with the jar in a horizontal position, shake and roll it gently to and fro, with an occasional sidewise movement, for about ten minutes. Then pour the contents into a pan and pan off the barren gravel and black sand. Save the mercury and amalgam as before. In this process, a small pinch of potassium cyanide is better than the teaspoonful of lye. However, since potassium cyanide is extremely poisonous, its use is not recommended unless you are fully prepared to observe the precautions entailed by its possession and use.

Somewhere above I mentioned that one of the valuable characteristics of quicksilver was its high specific gravity which permitted it to lie quietly under a moving stream of water and placer gravel. Hence, add some mercury carefully to your sluice boxes where it will work its way in and around the riffles to form little pools. Do not add quicksilver to the sluice boxes until the water and gravel have run through long enough to cause the lumber to swell sufficiently to seal all cracks. Otherwise, you'll lose quicksilver — and since it costs several dollars per pound, you stand to lose some money! At best, you'll lose from five to ten percent of the mercury anyway. However, the increased gold recovery resulting from the use of quicksilver ordinarily more than compensates for the small loss of quicksilver. Here are a few tips: Until you have some experience, add only a few pounds of quicksilver to each sluice box section; have a light stream of water flowing through the sluice while you add the quick; hold the quicksilver container close to the water while you pour the quick into the sluice so as to minimize both flouring and exposure to air; and use quick in only the first few boxes of the sluice.

After having obtained your gold amalgam from your panning or sluice box concentrates, the problem arises as to what should be done with it. A considerable amount of mercury in the amalgam can be removed by simply placing the amalgam in a piece of chamois skin (the same type you use to clean the windshield and other glass in your automobile) and squeezing out the excess quicksilver. It is best to do this under water. The excess mercury thus removed will contain not to exceed about 1/10 of 1% of gold and is excellent to use again in your pan or sluice boxes. You now have left a quantity of thick, pasty amalgam which contains anywhere from 25% to 55% of gold and silver.

I could tell you to retort your amalgam, thus saving the quicksilver, and to send the retort sponge (the gold) off to the mint. However, my experience in observing the practice of beginning operators leads me to conclude that retorting had best be deferred until you have had an opportunity to watch an experienced operator. Otherwise, you run an excellent chance either of salivating yourself (mercury poisoning) or of blowing yourself up (due either to clogging of the retort or to the sucking of water up through the condensing tube into the hot retort.) Should you have only moderate amounts of amalgam, the loss of the contained quicksilver is not of much importance. In that event, you can place your ball of amalgam on any clean iron surface over an open fire. The heat drives off the mercury as a vapor, which may or may not be visible, and leaves the gold. The gold may be sold without further treatment. In following this plan, it is a good idea to cover the amalgam with a small inverted tin can in which a few very small holes have been punched. Should the amalgam tend to "spit" and disintegrate when heated, small flying particles of gold will not be lost. Another thing, the whole procedure should be conducted well out in the open so that you will not inhale any of the mercurial vapor — it's poisonous!

Another plan which has certain advantages is to place your lumps of amalgam in a jar containing an acid solution consisting of one part of concentrated nitric acid and one part of water. Incidentally, distilled water is preferable to ordinary water in that the former will contain no chlorides. The acid will dissolve the mercury, leaving the gold either as a spongy mass or as a fine dust. After the mercury is dissolved, pour off and save the solution. Carefully wash the gold several times with water, dry, and it is ready to be sold. The solution you poured off is excellent to amalgamate the copper bottom of your pan. Just pour it over the bottom, let stand for a minute, and pour the excess back into the bottle. Rub the bottom with a clean cloth and there you have a bright amalgamated surface ready to trap more gold for you!

CHAPTER IX

WANTED: A PLACER CLAIM

Although prospecting for placer gold is a lot of fun, it usually entails a lot of work and quite a bit of time. Even after you have worked diligently and long, it is not predetermined that you shall locate a good gold placer. Frequently, the prospective gold-panner is interested in the field only as a fascinating hobby to be indulged in on week ends or vacation periods. Or it may be that you have retired but would like a healthful, outdoor activity that you can pursue as you see fit, but which you would like to have pay its own way.

In any event, either the leasing or outright purchase of a small placer deposit may prove to be your best bet. Should you lease a property, you will find you have somewhat less freedom of action than if you purchase a claim. Leases follow local custom more or less, but usually run for a period of from one to five years, with or without option of renewal. Leases generally provide for a royalty payment of ten to fifteen percent. However, a lease usually requires that you, the lessee, perform a regular amount of work each month. If placering is to be a part-time proposition, the terms of a lease might be pretty hard to observe.

In some areas you may find folks who own placer ground but who are not personally interested in working it. Often an informal agreement may be reached whereby you may be permitted to do small scale placering, either with or without payment of a royalty to the owner. During the depression of the thirties, a great many people worked various privately owned placer deposits, in many cases without even going to the trouble of first obtaining permission. I do not recommend such promiscuous "sniping" (as it is called), since permission to use your pan or sluice is ordinarily not too difficult to obtain. But do not jump to conclusions — such "free" placer deposits are usually not very rich, and may not prove to be much more than hard work.

The outright purchase of a placer claim or mine is a different proposition. In such a case, the real merit of the property first should be demonstrated to your satisfaction. After purchasing the mine, you can work it as little or as much as you like provided it is patented. On the other hand, if the claim is unpatented, you are required to perform annual assessment work in the amount of one hundred dollars worth. By performing such annual assessment work, the claim may be held indefinitely. Or, after the performance of annual assessment work for five years, you can apply for a patent, after the granting of which no further assessment work is required.

At the time of writing this book, there are quite a large number of claims available for purchase. Some of these claims are offered directly by private owners; others are offered by licensed mineral brokers. In general, by writing or personally contacting such a mineral broker, you will be able to determine quickly what properties are available, their locations, conditions, and prices. The best way to get an idea of the merits of any gold placer which may be offered to you is to inspect it thoroughly in the company of the owner or his agent. Have it thoroughly demonstrated that the placer is actually capable of producing placer gold and, presumably, in paying quantities. Be sure there is an adequate water supply for at least a portion of the year. Be equally certain that you have adequate disposal facilities for the tailings. As to the legal aspects of buying a placer claim, my considered advice is to have the final details handled by an attorney. His fee will be nominal and his services may help considerably in protecting your investment.

The current price of placer claims (unpatented), usually of twenty acres, varies from a minimum of around $500 to a maximum which depends upon the size and richness of the claim.

CHAPTER X

THE LAW SAYS

All valuable mineral deposits in land belonging to the United States both surveyed and unsurveyed, and the lands in which they are found, are free and open to exploration, occupation, and purchase, except that aliens may not obtain patent. This includes mineral land within a national forest, the unpatented parts of congressional grant to a railroad (except coal and iron), or of an unconfirmed Mexican grant.

Generally speaking, land is not subject to mineral location when lying within a subsisting Indian, military or naval reservation; national parks, national monuments; the bed of a navigable river, privately owned land, and patented land deeded to the state for unpaid taxes. Other land is also not subject to mineral location, depending upon the particular state involved. When in doubt, write to the Division of Mines or Department of Natural Resources of the particular state in which you are interested.

The law permits the location of mining claims within the boundaries of national forests. I say, the law permits it. But from a practical standpoint, you'll probably find that the Forest Service will view your claim with a dubious eye. You may be able to patent your claim located in a national forest, but you'll probably have to battle the Forest Service every inch of the way. Such fighting is conducted in the court room and you will be required to present ample proof that your land is more valuable for mining than for any other purpose. Hence, my humble advice is to forget prospecting in national forests. Should you still want to prospect in a national forest, I then advise you not to locate and attemppt to patent any placer claim therein unless it is extraordinarily rich. May I point out that the Forest Service is not so much opposed to mining as it is interested in fire prevention, erosion control, improvement of the forests and range land, etc. In furthering these meritorious aims, it is only natural that the Forest Service might come in conflict with the prospector and miner.

In locating a placer claim, the essential elements are the discovery of minerals, marking of boundaries, posting the notice of location, and the recording of an exact copy of the notice. Your placer location notice positively must contain the following information: (1) date (2) name of the locator, or locators, (3) name of the claim, (4) the number of feet or acreage claimed, and (5) the distance and direction, as accurately as possible, from the discovery point to some permanent and well known point, natural object, or monument. When your placer claim is located and described in the notice by legal sub-divisions, no other tie-in is required. Post your location notice at the discovery point and protect it from the elements by placing it in an inverted box or tin can, and in plain view.

The maximum size of a placer location is 20 acres for an individual and 160 acres for an association of not less than eight persons, or if the association is composed of less number, 20 acres for each individual therein. The smallest placer location recognized by Federal law is 10 acres, which must be square in form.

A location may be made without regard to the age, sex, residence, or citizenship of the locator.

The foregoing is presented to give the beginner placer miner an idea of the formalities involved in "staking a claim." But it should be distinctly understood that what is set forth above is not complete. The recording of the claim, performance of discovery labor, and the performance of annual assessment work, to mention just a few items, are all extremely important and must be followed to the letter. A complete description of all these factors would be out of place in this little handbook. Accordingly, I have a very complete booklet on how to locate and hold mining claims which I shall be pleased to send you upon receipt of only $1.50 (in coin or stamps). Your claim may be worth a great deal of money when properly located, but it isn't worth a cent if an improper location has been made. The booklet I mentioned above is the best claim insurance I know of and I sincerely urge that you send for your copy now. Claim location blanks will also be found advertised in the back of this book.

CHAPTER XI

HOW TO SELL YOUR GOLD

If you have worked diligently, and have had even a modest amount of luck, you should quickly arrive at the point where you have a nice amount of placer gold to sell. In general, there are five classes of buyers for your gold: (1) Individual gold buyers, (2) Local stores, (3) Local banks, (4) Smelting companies, and (5) United States mints and assay offices. In the course of ordinary small placer operations, you probably will have no occasion to sell to smelters.

If you have only a small amount of gold for sale, you may find it best to sell to one of the local stores who purchase gold as a convenience for their customers. Naturally, the merchant will discount the gold one or two dollars an ounce, and since he can only estimate the fineness of the gold, he stays on the safe (i. e., lower) side. However, the above small disadvantages may be offset by the fact that you not only receive immediate cash for your gold, but by the fact the merchants will usually accept lots of less than one ounce — the minimum quantity accepted by the mints.

When you have accumulated a sufficient quantity of gold, you will probably do best to send it to a United States mint. You can send your gold either by mail or by express — in either case, see that it is fully insured. When you send a shipment to the Mint, send a separate letter in which you have enclosed the necessary affidavit. If you mined the gold yourself, the necessary form is TG-19 which you can obtain by addressing a request therefor to any United States mint or assay office. If the amount of gold in the shipment is 5 troy ounces or less, the form need not be notarized.

At the present time the mints are paying $42.22 per troy ounce of fine gold, subject to certain charges. The mint charges one dollar for melting any deposit of 1,000 ounces or less and ten cents additional for each 100 ounces over this amount. An extra charge of one dollar or more is made for melting gold dust or gold containing non-metalics if the loss in weight in melting is more than 25%. If the gold is 992 fine, or finer, no charge is made for parting and refining. If less fine, or if more than 50 parts base metal per 1,000 are present, charges of one to five cents or more per ounce are imposed for parting and refining. Bullion less than 200 fine is not accepted.

Current prices are paid for the full silver content. However, if necessary forms are submitted for silver qualified under Executive Proclamation of December 21, 1933, the depositor will receive the number of silver dollars that can be coined from one half of the fine silver content. No other constituent of the bullion is paid for.

From the foregoing, it is apparent there are two major factors that govern the price you receive for your gold, i. e., the fineness and the weight. At first, it will probably be pretty difficult for you to judge the fineness of your gold. Of course, an assay can be run to determine the fineness, but that is practically a waste of money. When you ship your gold to the mint, there is no recourse from the mint assay returns whether you like them or not. After you have made a few shipments to the mint of gold mined in various localities, you will develop the ability to pretty accurately judge the fineness of any particular lot of gold.

It is no great problem to weigh your gold. Although you can buy balances and weights for from forty dollars and upward, you can easily make a set of hand balances that will weigh one or two ounces of gold to within about a grain, or the nearest five cents worth.

The balance may be made of wood, six or eight inches long, suspended by a pin, needle, or bent wire hook through a small hole in the exact middle of the beam. The pans can be made of tin, cut 1½ or 2 inches in diameter, and hammered dish-shaped, or the pans can be made from the lids of small tin cans, each suspended by three threads from the ends of the beam by means of bent wire hooks. No pointer is necessary as the beam can be levelled closely enough by eye. It is not necessary that all parts be of exact weight, as the balance beam or the pans can be trimmed to make the assembly balance. For even approximate accuracy, however, it is necessary that the pans be suspended from points on the beam the same distance from the center bearing, and for stability the end bearings must be slightly lower than the center one. The nicer the construction and the more nearly frictionless the method of suspension, the greater the accuracy. Even with very little attention to these points, a sensitivity of less than a grain is obtainable when weighing as much as two ounces.

Although weights can be purchased in convenient sets for about four dollars, you can easily improvise your own. In the back of this book you will find a table that shows the weights of coins of various denominations from one penny to one silver dollar. By using various combinations of coins, you will be able to do a very practical job of weighing your gold.

Please refer to item No. 82 on page 48. It's darned good!

CHAPTER XII

YOUR PROSPECTING OUTFIT AND SUPPLIES

An automobile is to be preferred for transportation in regions where roads and conditions permit its use. Otherwise, the prospector may use one or two burros. The burro can pack about 150 pounds and live off the country. The automobile permits more complete outfits to be carried and trips out for supplies can be made easily and quickly.

A minimum camp outfit should contain a roll of three or four blankets (depending upon the climate) in a canvas cover, a tent, a tight wooden box with lid for storing food away from insects and rodents, and a canvas "war bag" for clothes. A stove will be needed when prospecting is done in cold weather. In addition, a folding cot will be quite desirable.

A full size ax and a stout pocket knife are absolutely essential. A saw, hammer, and a few pounds of assorted nails will come in handy in fixing up a camp. A 50-foot length of manila rope will be found useful. A miner's acetylene lamp and a 5-pound can of carbide will furnish a good light and last a camp all summer. A flashlight and a supply of batteries will be a worthwhile convenience. A 2-quart canteen with a shoulder strap usually is needed for carrying drinking water or water for panning. Under hot weather conditions, a 10-gallon tank of water will last one man three days for drinking and cooking. In cooler weather, 10 gallons of water should last a man a week or ten days. This does not make allowance for the radiator of a car.

A pick, a long-handle round pointed shovel, a gold pan, and a prospector's pick are required. A compass will be useful in running lines if claims are to be staked. A hand magnifying glass is a great help in identifying minerals, and a mortar and pestle and small pan will be needed in testing rock for free gold.

The Gold Testing Outfit advertised in the back of this booklet has already proved its worth to countless prospectors and should be included in every prospecting outfit. A single-jack hammer with two or three moils (or cold chisels) are required for taking samples and for loosening rock in making cuts. Some prospectors carry one or two sets of hand steel and several pounds of powder. A few rounds may be drilled and blasted before the steel has to be resharpened.

A one or two man party will need a frying pan, coffee pot, a large and a small stew pot or pan, and a Dutch oven. A knife, fork, spoon, cup, and plate should be provided for each man. Bring along a good butcher knife, water pail, can opener, and a few tea towels. Bacon, flour, beans, oatmeal, canned or dried fruit, coffee, syrup for hot cakes, and sugar and canned milk for coffee are the standbys in prospectors' camps. Canned tomatoes are in common use and they are a splendid aid in helping to prevent scurvy. Fresh meats are difficult to keep and their attempted use in summer prospecting camps is not recommended. The following will provide a balanced diet for one week for one person:

> Three 1-lb. cans evaporated milk, 2-lbs. potatoes, 4-lbs. onions, cabbage, beets, or other vegetables, 3-lbs. citrus fruits, 6-lbs. fresh apples, or equivalent in dried prunes, apricots, etc., 3-lbs. dried beans; 6 to 8-lbs. cereals, whole wheat flour or bread, rolled oats, shredded wheat, etc.; 2½-lbs. dried meat, bacon, ham or cheese, (fresh meat or eggs may be substituted if available); 3-lbs. sugar; 1-lb. coffee; 1/4-lb. butter, and baking powder, salt, etc.

A few years ago the prospector could get by with about 50c a day for provisions. With today's costs, not less than a dollar a day should be allowed.

Don't forget that your most important item of clothing should be a pair of stout, thick-soled shoes, of good quality, and preferably hob-nailed. Woolen socks worn under the heavy boots will help to prevent blisters; you will wear out several pairs of these socks in a season. Take along other clothing according to the climate and service. In mountainous regions, flannel shirts, woolen underwear, and overalls are usually worn. No matter what you take along, at least be sure that you have one complete change of dry clothing just in case you get caught in the rain or have to work in the rain or dampness all day.

Finally, since you probably won't be close to medical aid, some first aid and medical supplies should be included in your outfit. As a minimum, these supplies should include a laxative, iodine or merthiolate to disinfect cuts and bruises, some lotion or salve for sunburn or chapped lips, and a first aid kit. A snake bite kit (not whisky!) may also prove very valuable.

CHAPTER XIII

GOLD NUGGETS FOR SALE

My stock of genuine gold nuggets is the finest commercial collection in the United States. For nearly twenty years I have been acquiring some of the choicest gold nuggets mined in the West and in Alaska. Now I am in a position to furnish my customers with just about anything they may need in the form of gold nuggets.

These very choice gold nuggets are especially suitable for having your jeweler (or yourself) use in earrings, brooches, tie pins, cuff links, watch chains, necklace pendants, etc. Gold nugget jewelry makes a gift that is both beautiful and distinctive. Gold nuggets make exquisite birthday and anniversary gifts, too. The most exciting gift you could possibly make to any rock collector (old or young) is a genuine gold nugget.

Since most folks aren't very familiar with sizes and weights of gold nuggets, let me give you an idea or two. Their unit of weight is the PENNYWEIGHT. There are 20 pennyweights (abbreviated dwt.) in a troy ounce; the troy ounce is somewhat heavier than the usual commercial ounce. A pair of nice nuggets weighing a total of 1 or 1½ dwt. would do nicely for earrings or cuff links. Or you could use a cluster of 2 or 3 smaller nuggets, but with the same total weight, for each earring or cuff link. A ladies brooch could use perhaps 6 or 8 small nuggets, weighing a total of not over 2 dwt. A pendant for a necklace should be a very smooth nugget running from 1 to not over 2 pennyweights. A man's tie or "stick" pin can be either a smooth or rugged nugget weighing around 2 dwt. A nugget to hang from a watch chain should weigh between 2 and 4 dwt. For a pocket piece, the nugget can go up to 20 dwts. (If you go over that it's downright uncomfortable to carry!) Recently I have seen a number of most strikingly beautiful pieces of jewelry made of clear plastic with bright gold nuggets embedded therein.

My nuggets are the choicest to be found anywhere. And whether you order tiny nuggets, or a whopping big one, the quality will be the same. They price you pay is based strictly on weight. On small orders, the price is $4.00 per pennyweight (dwt.)

Following is my schedule of prices:

Pennyweights Ordered	Price per Pennyweight
Less than 3	$4.00
3 up to 5	$3.50
Over 5	$3.00

Naturally, your best bargain will be an order over 5 pennyweights. The above prices are postpaid and include full insurance.

When sending in your orders, be sure to tell me how many nuggets you want. And if you want them for any special purpose (such as jewelry), let me know so I may personally select the very best for you. If you're in a real hurry, include an air mail stamp or two, and I'll personally get your order off via air mail the same day I receive it.

Following are a few of the gold nugget items that have proved very popular and which you'll like, too:

GOLD 1: One nice shiny gold nugget $1.00

GOLD 2. Three nuggets similar to above 2.00

GOLD 3: A choice collection of 12 nice gold nuggets 5.00

GOLD 4: A big bright chunky nugget that will amaze you! A dandy pocket piece! 10.00

GOLD 5: My '49er Collection! A most generous collection of lovely nuggets. Many odd and unusual shapes—but just as they come from the mines. You will be very pleased 25.00

Send your order to:

WALTER J. ROBERTSON

P. O. Box 604 Stockton, California 95201

BIBLIOGRAPHY

GENERAL

Jackson, C. F., and Knaebel, J. B., Small-Scale Placer Mining Methods; U. S. Bureau of Mines I. C. 6611 (1932).

Gardner, E. D., and Johnson, C. H., Placer Mining in the Western United States — Part I, General Information, Hand-Shoveling, and Ground Sluicing; U. S. Bureau of Mines I. C. 6786, (1934).

Gardner, E. D., and Johnson, C. H., Placer Mining in the Western United States — Part II, Hydraulicking, Treatment of Placer Concentrates and Marketing of Gold; U. S. Bureau of Mines I. C. 6787 (1934).

Placer Mining Methods; Report of Investigations 2315; U. S. Bureau of Mines (1922).

ALASKA

Wimmler, N. L., Placer Mining Methods and Costs in Alaska; U. S. Bureau of Mines Bulletin 259 (1927).

Brooks, A. H., The Mineral Deposits of Alaska; Mineral Resources of Alaska; U. S. Geological Survey, (1913-1914).

Hutchins, H. P., Prospecting and Mining Gold Placers in Alaska; U. S. Geological Survey Bulletin 345.

Purington, C. W., Methods and Costs of Gravel and Placer Mining in Alaska; U. S. Geological Survey Bulletin 263 (1905).

Gold Placers of Parts of Seward Peninsula, Alaska; Bulletin 328; U. S. Geological Survey, (1908).

Platinum — Hints for Prospectors; Pamphlet 34549; U. S. Geological Survey.

CALIFORNIA

Haley, C. S., Gold Placers of California; California State Mining Bureau Bulletin 92 (1923).

Laisure, C. M. and others, Mining In California; Reports of the State Mineralogist, Vol. 28, No. 2, Division of Mines, San Francisco.

Winston, W. B., and Janin, C., Gold Dredging in California; Bulletin 57; California State Division of Mines (1910).

Horner, R. R., Notes on the Black Sand Deposits of Southern Oregon and Northern California; Technical Paper 196; U. S. Bureau of Mines (1918).

Logan, C. A., Platinum and Allied Metals in California; Bulletin 85; California State Division of Mines (1919).

ARIZONA

Wilson, F. D., and Tenney, J. B., Arizona Gold Placers and Placering; University of Arizona Bulletin No. 132; Arizona Bureau of Mines, Mineral Technology Series No. 34; Tucson, Arizona.

Fansett, Geo. R., Small-Scale Gold Placering; University of Arizona Bulletin, Vol. 3, No. 1, January 1, 1932.

BRITISH COLUMBIA

Galloway, J. D., Placer Mining In British Columbia; Department of Mines, Victoria, B. C. (1931).

IDAHO

Finch, J. W., Prospecting for Gold Ores; Pamphlet No. 37; Idaho Bureau of Mines & Geology, Moscow, Idaho (1932).

Fahrenwald, A. W., Recovery of Gold From Its Ores; Pamphlet No. 37; Idaho Bureau of Mines and Geology (1932).

Staley, W. W., Elementary Methods of Placer Mining; Pamphlet No. 35; Idaho Bureau of Mines & Geology (13th edition, 1944).

Hill, J. M. Notes on the Fine Gold of the Snake River Idaho; Bulletin 620-L; U. S. Geological Survey (1916).

MONTANA

Dingman, O. A., Placer Mining Possibilities in Montana; Bureau of Mines & Geology Memoir No. 5; Butte, Montana (1932).

Jennings, H., The History and Development of Gold Dredging in Montana, with a chapter on Placer Mining Methods and Operating Costs by C. Janin; U. S. Bureau of Mines Bulletin 121 (1916).

NEVADA

Vanderburg, W. O., and Smith, A. M., Placer Mining in Nevada; Vol. 30, No. 4 State Bureau of Mines, Reno, Nevada (1936).

NEW MEXICO

Wells, E. H., and Wooten, T. P., Gold Mining and Gold Deposits in New Mexico; Circular No. 5, New Mexico School of Mines, Socorro, New Mexico.

OREGON

Pardee, J. F., Beach Placers of the Oregon Coast; Circular 8; U. S. Geological Survey (1934).

Horner, R. R., Notes on the Black Sand Deposits of Southern Oregon and Northern California; Technical Paper 196; U. S. Bureau of Mines (1918).

SOUTH DAKOTA

Anderson, D. L. M., Prospecting For Placer Gold in South Dakota; State Geological Survey; University of South Dakota, Vermillion, South Dakota (1933).

CONNOLLY, J. P., Geologic History of Black Hills Gold Placers; State Geological Survey; University of South Dakota, Vermillion, South Dakota (1933).

WASHINGTON

Ingersol G. E., The W. S. C. Placer Mill; Mines Information Bureau Circular No. 2; Washington State College, Pullman, Wash.

Ingersol, G. E., Small Scale Methods of Placer Mining and Placer Mining District of Washington & Oregon; School of Mines, State College of Washington, Vol. 15, No. 6, (1933).

NOTE: I do not carry the publications listed in the Bibliography. They usually may be consulted in your public library.

APPENDIX

WEIGHTS AND MEASURES

Troy Weight

24 grains	1 pennyweight
20 pennyweights	1 ounce
12 ounces	1 pound

Equivalents

1 ounce troy *equals* 1.097 ounce avoirdupois *equals* 31.103 grams

1 ounce avoirdupois *equals* 0.911 ounce troy *equals* 28.35 grams

1 pound troy *equals* 0.8229 pound avoirdupois *equals* 13.166 ozs. avoirdupois *equals* 373.2 grams

1 pound avoirdupois *equals* 1.215 pounds troy *equals* 14.58 ounces troy *equals* 453.6 grams

The following table is based upon NEW coins only; gold at $35 per oz.

	Weight		Value of Equal Weight of Gold			
Coin	Grains	Pennyweights	1000 fine	900 fine	800 fine	700 fine
Copper cent	48.000	2.000	$ 3.50	$ 3.15	$ 2.80	$ 2.45
Nickel	77.160	3.215	5.63	5.06	4.50	3.94
Dime	38.581	1.608	2.81	2.53	2.25	1.97
Quarter	96.452	4.019	7.03	6.33	5.63	4.92
Half dollar	192.904	8.038	14.07	12.66	11.25	9.85
Silver dollar	412.500	17.188	30.08	27.07	24.06	21.06

The average fineness of all placer gold mined in Nevada between 1906 and 1934 was approximately 713.

YOU ARE INVITED TO JOIN

AMERICAN PROSPECTORS

IT'S NEW! IT'S INTERESTING! IT'S EXCITING!

Membership Includes the Following Privileges:

1. Pan gold on our rich Journey's End gold placer claim in the scenic Feather River country of California. You'll love it! Keep the nuggets you find!
2. 10% discount on anything, including gold nuggets, advertised in this booklet.
3. Use of our rental Prospector's Library — a large collection of books on mining, prospecting, geology, mineralogy, etc.

We'll shortly start publishing our own magazine,

"GOLD NUGGETS"

only 25c a copy to members

MEMBERSHIP CONSISTS OF TWO CLASSES:

1. REGULAR MEMBERS — Dues are $1.00 per year.
2. LIFE MEMBERS — Costs $10.00 to join, but no annual dues thereafter. Life members receive a magnificent membership card with a lovely big gold nugget mounted on it, and the card with nugget covered with clear plastic.

You'll be mighty proud to show this card to your friends.

All our present members will be very glad to have you join them. It's very simple.

Just mail your name, address, and membership fee

($1.00 to become a regular member;

$10.00 for life membership) to

American Prospectors

P. O. Box 604
Stockton California 95201

No. 1—16-inch MINERS' PAN. Beautifully made of polished steel, with rolled edge for reinforcement. Designed to give you complete satisfaction. .... **$ 3.00**

No. 2—12-inch MINERS' PAN. Same as above except somewhat smaller. .... **$ 2.00**

No. 4—THE DOODLE BUG blueprint, a wet concentrator used in connection with a gasoline engine, pump, and hose .... **$ 1.00**

No. 5—THE PAPOOSE (Rocking Sluice) blueprint, a wet concentrator—handles 50 times as much dirt as a pan .... **$ 1.00**

No. 6—THE HAND DRY WASHER blueprint. A dry concentrator, which weighs about 32 pounds, for use where water is not available .... **$ 1.00**

No. 7—THE POWER DRY WASHER, 2 blueprints. A dry concentrator, weighs about 50 pounds, uses 2-cycle gasoline engine, and will keep one man busy shoveling .... **$ 1.50**

No. 8—THE SLUICE BOXES blueprint. Shows different kinds of sluice boxes, how to set them up, and the kinds of screen and riffles to use for best results .... **$ 1.00**

No. 9—THE SAMPLER DRY WASHER blueprint. For an 8-pound dry concentrator which you can fold and carry under your arm! .... **$ 1.00**

No. 11—Any three of the above blueprints. .... **$ 2.00**

No. 12—CLAIM LOCATION BLANKS. Always carry a supply with you. It requires three per claim: One left in the monument, one left at the court house, and one for your own record. In ordering, state whether for LODE or PLACER claims. 3 for .... **$ .40**

No. 13—CLAIM LOCATION BLANKS. 9 for .... **$ 1.00**

No. 14—BLACK SAND & GOLD. For those who wish to practice panning with genuine black sand and gold particles. Taken from actual cleanup. Per package. .... **$ 1.00**

No. 15—HOW TO LOCATE & HOLD MINING CLAIMS, new edition. (Also applies to uranium claims, as well as gold, etc.) .... **$ 1.50**

No. 17—GENUINE CALIFORNIA PLACER GOLD! This vial of real placer gold, recovered from the placer diggin's of California's world famous Mother Lode will give you a real thrill and in all likelihood a bad case of "gold fever." Only one vial to a customer. .... **$ 5.00**

No. 18—Jumbo-size Postcard showing gold in Natural Color! Measures 6⅝"x9⅛". Shows a collection of prize winning specimen gold from Siskiyou County, California—placer and quartz nuggets. Only known collection of its kind in the world: shown at World's Fair, Treasure Island. Each .... **$ .25**

No. 19—FIELD TEST SMELTER! Consists of one crucible and enough powder for twenty tests. Equal amounts of ore and powder are mixed together and placed in crucible, and then ignited with a match. The resultant terrific heat (about 1,400 degrees) releases the metals from their chemical combination in the ore, and they form small globules in the bottom of the crucible, where they can be readily identified! A wonderful prospecting aid for only .... **$ 3.00**

No. 20—PAY STREAKS ON THE BARS. This is a sheet of information on how to locate the different pay streaks on the river bars. How they are formed, their color and texture. Invaluable to the man working on the river bars. .... **$ .50**

No. 21—A NICE PLACER GOLD NUGGET. Exactly as recovered in actual mining here in California. If you would like to have a nugget of genuine native gold, this is it! .... **$ 1.00**

No. 22—PROSPECTOR'S MAGNIFYING GLASS. A powerful magnifying glass. Very useful for careful examination of your minerals, gold nuggets, etc. Just imagine, only .... **$ 1.00**

No. 29—GOLD TESTING OUTFIT. Can be used to quickly test rock or ore for gold in just a few minutes. The test is simple, yet reliable. The outfit is packed in a sturdy box and is easy to carry anywhere. No running water, electricity, etc., are required—hence you can use the outfit up in the hills, along the creeks, out in the desert—in fact, just about anywhere! Shipped with complete instructions, postpaid, and insured. .... **$ 6.00**

No. 30—GOLD AND SILVER TESTING OUTFIT. As above, except you can also test for silver. .... **$ 7.00**

No. 34—"FIELD TESTS FOR ELEVEN VALUABLE METALS." This interesting pamphlet tells you how to make tests right in the field for barium, chromium, copper, gold, lead, manganese, mercury, nickel, silver, tungsten and zinc. Every prospector needs a copy. .... **$ 1.00**

No. 37—ART OF GEM CUTTING—Dake. New Sixth Edition. Completely illustrated. Most complete book available on gem cutting for amateur and commercial gem cutting. Covers both cabochon and facet cutting. .... **$ 2.00**

No. 38—"MINERAL RESOURCES OF COLORADO." 547-pages, with hard covers. A superb collection of information covering metals, non-metals, and fuels in all known mining deposits and all mining districts of the state. A pocket in the back of the book contains a large multi-colored map showing precise locations of Colorado's metallic mineral deposits. This book is now scarce and my copies are limited. If you're interested in Colorado, order at once. Postpaid. .... **$ 4.00**

No. 39—GEOLOGY BOOKS. A few evenings spent with a good geology book will teach you what it took many of us years to learn the hard way. These books are used, but guaranteed in good condition. Most over 300 pages, and beautifully illustrated with photographs, maps, charts, etc. Per copy, postpaid. .... **$ 2.00**

No. 42—URANIUM AND FLUORESCENT MINERALS, Dake. Prospecting and detection of uranium. Using the Geiger counter. Locations where found. .... $ 2.00

No. 45—BLOWPIPE. Made of polished brass, 8-inches long. Excellent for use in mineralogical determinations. A hard to get item. .... $ 1.50

No. 61—A MODERN METAL LOCATOR. Instructions and diagrams for building an economical and sensitive device for locating metal ores, buried treasure, etc. Build your own and save money. .... $ 1.00

No. 62—PROSPECTING FOR GOLD, by Idriess. Completely covers both lode and placer gold. Includes full data for the cyanide process, too. Easy to read and understand; fine for the beginner prospector. 296 pages, hard covers. .... $ 4.00

No. 63—A SIMPLE GEIGER COUNTER. Instructions and diagram for an amazingly simple Geiger counter of excellent sensitivity. Can be built very cheaply. .... $ .50

No. 64—WHERE TO LOOK FOR URANIUM IN CALIFORNIA, by Raymond. Complete data on where you may look, what to take, what to look for, how to file a claim, together with tables, forms, and maps. .... $ .50

No. 65—FORTUNES IN MINERALS, by Idriess. The finest book of its kind on the market. It often has been referred to as the Prospector's Bible. With the instructions contained in this book you can identify just about any commercial metal or mineral you find. Its description of simple field tests is a feature to be found in no other book. Its treatment of uranium is very good. 310 pages, hard covers, postpaid. .... $ 6.00

No. 66—CONFIDENTIAL TIPS ON GOLD PLACERING IN NEVADA, by Robertson. Not a map, but the real "low-down" on where to go, what to expect, the best spots, etc. For either wet or dry placering, this is what you need. .... $ 1.00

No. 67—MINERALS OF CALIFORNIA, by Murdoch & Webb. Lists 523 mineral species found in California, together with their locations. 559 pages, 3 color plates, hard cover, and 2000-entry bibliography. A most valuable aid to the prospector, mineralogist, rockhound, etc. Limited edition. .... $ 7.00

No. 69—LOST MINES AND HIDDEN TREASURE, by Lovelace. The most fascinating book of its kind ever written. Covers Arizona, Colorado, California, Nevada, Utah, and Texas. A scholarly work that not only describes the mines and treasures, and how they were lost, but furnishes actual directions and clues for finding this fabulous wealth. Covers, gold, silvers, opals, platinum, and diamonds! Makes a wonderful gift for anyone. .... $ 4.95

No. 70—HOW TO KNOW THE MINERALS AND ROCKS, by Pearl. Gold, diamonds, turquoise, garnets, marble, meteorites—these are just a few of the beautiful and interesting minerals and rocks that you will learn to identify with the help of this handy, authoritative field guide. Includes 46 full color illustrations, 156 drawings, a total of 192 pages, soft covers. .... $ 1.00

No. 73—LET'S BUILD A SWIMMING POOL!, by Robertson, doesn't have a thing to do with panning gold! But it's a real-do-it-yourself book for the man who wants to build his own pool and save money. Well illustrated with sketches and photographs, soft covers. .... **$ .50**

No. 76—GOLD RUSH COUNTRY — GUIDE TO CALIFORNIA'S MOTHER LODE & NORTHERN MINES, a Sunset Discovery Book. This book is full of magnificent photographs, detailed maps, on large 8"x10¾" pages. The most thrilling book of its kind ever published. Postpaid. .... **$ 3.00**

No. 77—NORTHERN CALIFORNIA, a Sunset Discovery Book. If you plan to come here, this book will give you an insight into the Northern California that is waiting to be explored. Profusely illustrated, large 8"x10¾" pages. Postpaid. .... **$ 3.00**

No. 78—SOUTHERN CALIFORNIA, a Sunset Discovery Book. Similar to No. 77 above, except the book covers the southern portion of the state. Postpaid. .... **$ 3.00**

No. 79—GEM HUNTER'S GUIDE, by MacFall. Contains 81 maps. Tells you exactly how and where to find precious gemstones. Includes clear photos of 53 gems in natural color. Lists over 800 locations over the United States where gem materials have been found. Full data on how to identify over 75 different gem materials. A magnificent book. Postpaid. .... **$ 6.95**

No. 82—PROSPECTOR'S GOLD SCALES. These very practical scales, complete with accurate weights, will do a good job for you. Not fancy, but rugged, sensitive and light-weight. A bargain! .... **$ 8.00**

No. 83—CONFIDENTIAL TIPS ON GOLD PLACERING IN ARIZONA, by Robertson. Where to go, what to expect, the best spots, etc. With original sketch maps. The very best available. .... **$ 1.00**

No. 84—GHOST GOLD, by Arnold. About the Lost Dutchman Mine in the Superstition Mountains near Phoenix — how and where to look for it. .... **$ 3.00**

No. 85—APPALACHIAN GOLD, by Campbell. Covers both lode and placer gold in Virginia, North Carolina, South Carolina, Georgia and Alabama, with maps and sketches. Real good. .... **$ 2.00**

No. 86—UNDERWATER PROSPECTING TECHNIQUES, by Greene. For underwater gold placering, this is the finest book available. Excellent discussion of equipment needed and available. You'll like it! .... **$ 2.00**

No. 87—CALIFORNIA GEM TRAILS, by Henry. 120 of the best localities. 25 pages of maps, 104 pages. .... **$ 2.50**

No. 90—NEW MEXICO GEM TRAILS, by Simpson. Exclusively for New Mexico with maps, directions, pictures and detailed information on rocks, gems, minerals, fossils and Indian artifacts. .... **$ 3.00**

No. 91—ARIZONA ROCK TRAILS, by Bitner. 20 maps with over 60 locations of interest. Maps of gem and mineral locations are printed on sturdy cards with descriptions and directions on back. Tells where to go, how to get there, and what to look for. .... **$ 2.00**

No. 92—THE ROCK COLLECTOR'S NEVADA & IDAHO, by Henry. Includes many maps showing exact locations of gem materials and mineral specimen areas. .... **$ 2.50**

No. 93—GEM TRAILS OF TEXAS, by Simpson, 40-full-page maps show 60 localities. 88 pages of detailed information as to the best localities in Texas. .... **$ 3.00**

No. 94—MIDWEST GEM TRAILS, by Zeitner. A field guide for the gem and mineral hunter in South Dakota, Illinois, Michigan, Minnesota, Ohio, Indiana, Nebraska, Kansas, North Dakota, Missouri, Iowa and Wisconsin. .... **$ 2.00**

No. 95—COLORADO GEM TRAILS, by Pearl. Detailed directions many mileage logs are given for the various mineral locations in Colorado. Hard bound. .... **$ 4.95**

No. 97—CONFIDENTIAL TIPS ON LODE AND PLACER GOLD IN THE SOUTHEASTERN STATES, by Robertson. Years have been spent in preparing this exciting list of hundreds of locations, helpful hints, etc. There are millions still left here! .... **$ 1.00**

No. 98—"PROSPECTING IN CANADA", by Lang. Excellent for Canada, but its thorough detailed instructions are fully applicable to the United States. Well illustrated, 401 pages. A real Bargain! .... **$ 4.00**

No. 99—HOW AND WHERE TO PAN GOLD, by Winters. A dandy book for the beginner. Lots of maps and photos. .... **$ 2.00**

No. 101—GEM TRAILS OF ARIZONA, by Simpson. A new field guide on Arizona, with 40 maps, over 55 well described locations, and lots of pictures and diagrams. .... **$ 3.00**

No. 105—DESERT GEM TRAILS, by Strong. Beautiful coverage of the Mojave and Colorado Deserts of California, Nevada, and Arizona. Detailed maps. Postpaid. .... **$ 2.00**

No. 106—EASTERN GEM TRAILS, by Oles. Your complete guide to gem and mineral localities in New Jersey, Pennsylvania, Maryland, Washington, D.C., Virginia and North Carolina. Very detailed information. Postpaid. .... **$ 2.00**

No. 107—PLACER GOLD MINING SUCCESS EVALUATION TEST. Easily test your abilities, experience, and resources to determine your chances to be successful at placer gold mining. Very easy to use. Available about two months after publication. Postpaid. .... **$ 1.00**

No. 108—GOLD AND DIAMONDS IN INDIANA, by Blatchley. No kidding, Indiana has them. Postpaid. .... **$ .75**

MINERAL COLLECTIONS

No. 53—THE PROSPECTOR MINERAL COLLECTION. Fifty different mineral ores, each ¾" or larger. A collection of mineral specimens assembled from hundreds of mine examinations. Being the practical outcome of actual field work, this collection is valuable to beginners in their work or to one who prospects for pleasure. Instruction pamphlet included. Collection is shipped in card board box 9" x 12" postpaid. .... **$ 7.50**

No. 56—THE COLLECTOR'S COLLECTION. This collection is an extension of No. 53 and its scope includes many of the rarer minerals, several gem stones, etc. The samples are of distinctive higher quality. Packed in an 18" x 24" box, this collection is a dandy buy. **$45.00**

In addition to the items advertised herein, I also carry a complete line of miscellaneous prospecting supplies, etc. If you need anything not included here, write to me and I shall be glad to quote prices.

If you have any questions about prospecting or mining, just write me a letter and I shall be most happy to answer your questions. Please enclose a stamped envelope for my reply.

Additional copies of this booklet are available for one dollar per copy postpaid. Send cash, check, or money-order to the address below.

(California residents please include sales tax)

WALTER J. ROBERTSON

P. O. Box 604 **Stockton, California 95201**